Contents

Preface 2

Chapter 1: Origin of Life 3-11

1.1 Millèr-Urey Experiment

1.2 Who was Stanley Miller?

1.3 Harold Clayton Urey

1.4 Oparin and Haldane Theory

1.5 Coacervate

1.6 Primordial Soup

1.7 Reddi Experiment

1.8 Lazzaro Spallanzani

1.9 Discovery of Microorganisms

1.10 Fossil Evidence of Life

1.11 RNA World Hypothesis

1.12 What is Life?

Chapter 2: Biological Macromolecules 12-26

2.1 Protein

2.2 Primary Structure of Proteins

2.3 Secondary Structure of Proteins

2.4 Tertiary Structure of Proteins

2.5 Carbohydrates

2.6 Fatty Acids

2.7 Nucleic Acids

2.8 Water

References

Preface

The first chapter describes Miller-Urey experiment, Oparin and Haldane Theory, Reddi experiment, fossils evidence of life and RNA World hypothesis. The second chapter describes various biological macromolecules like carbohydrates, proteins, lipids, and nucleic acids.

Chapter 1

Origin Of Life

Introduction:

Scientists are searching the questions "How life originate on the Earth?". Earth arose more than 3.5 billion years ago. Life arise from non-living matter. Whe life evolved on our Earth, very little amount of oxygen are found in the atmosphere.

1.1 Miller-Urey Experiment:

Miller-Urey experiment was supported by Alexander Oparin and J.B.S Haldane's (1952).

Requirment:

i. Water (H_2O).

ii. Methane (CH_4).

iii. Ammonia (NH_3).

iv. Hydrogen (H_2).

v. Two flasks.

Miller-Urey experiment was conducted by Stanley L. Miller and Harold C. Urey. The experiment was conducted in 1953 at the University of Chicago.

All the chemicals (CH_4, NH_3, H_2) were sealed inside glass tubes and flasks. Glass tubes and flasks are connected together in a loop. One flasks contains half filled water and another contains a pair of electrods. Then, the water was heated. Vapour released from water. The water vapour moves freely into the second flasks. Miller and Urey mixed the gasses and electrically sparked the mixture of gasses to produce lighting. Lighting was stimulated. The gasses were condensed into liquid.

Miller identified five types of amino acids like Glycine, α – Alanine, β – Alanine, Aspartic acid, and α – aminobutyric acid.

Result:

This theory proved that first life arise spontaneously from inorganic materials. All the inorganic molecules are combined to produce organic compounds.

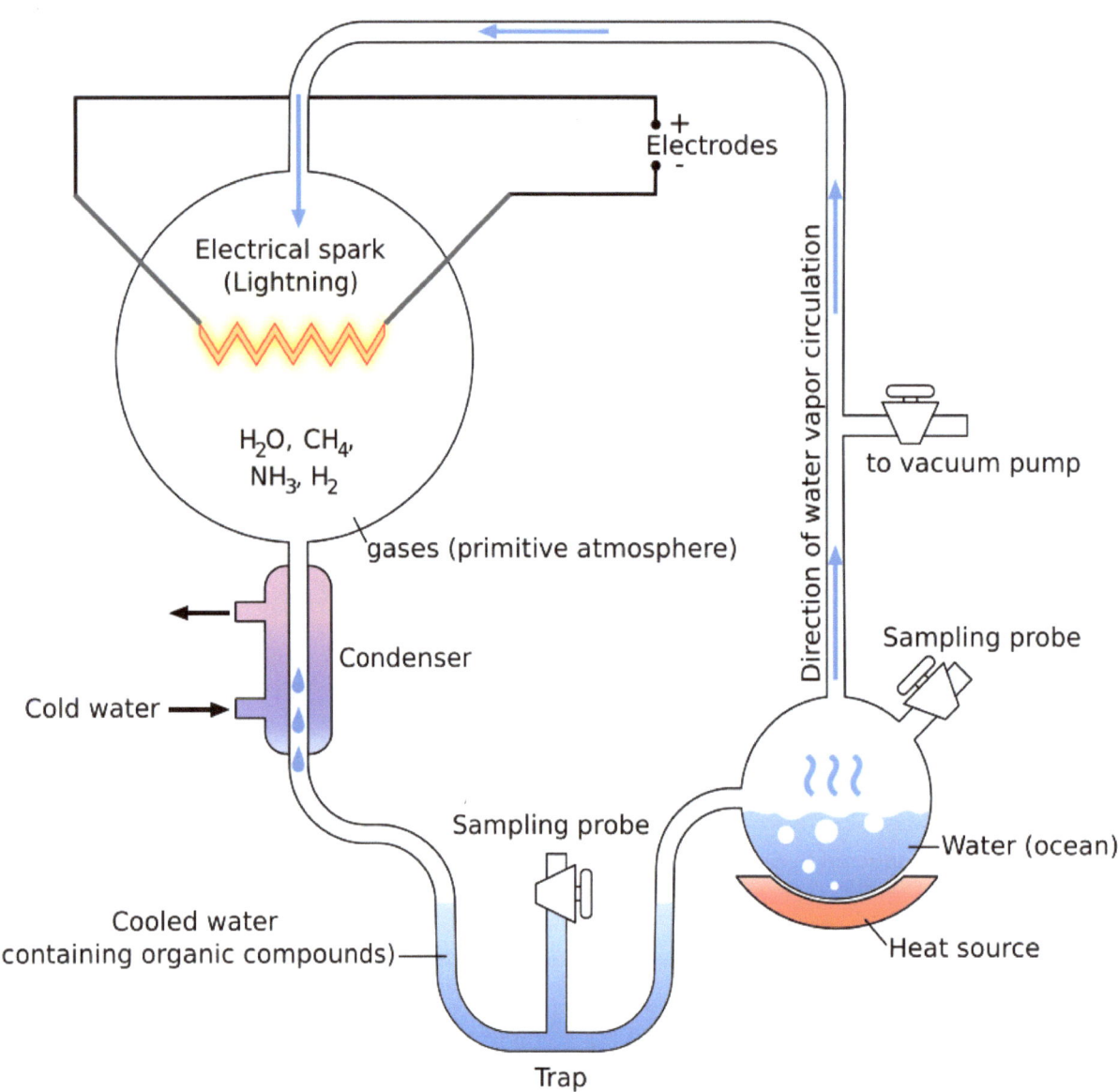

Figure 1. Miller-Urey Experiment (Photo Credit: Wikimedia Commons/GYassineMrabetTalk✉ This W3C-unspecified vector image was created with Inkscape. iThe source code of this SVG is valid. [CC BY-SA (https://creativecommons.org/licenses/by-sa/3.0)])

1.2 Who was Stanley Miller?

Stanley Miller was born on March 7, 1930. He was an American chemist. He was born in California, Oakland, United States. Miller was considered the "Father of Prebiotic Chemistry". Miller was died on 20 May, 2007.

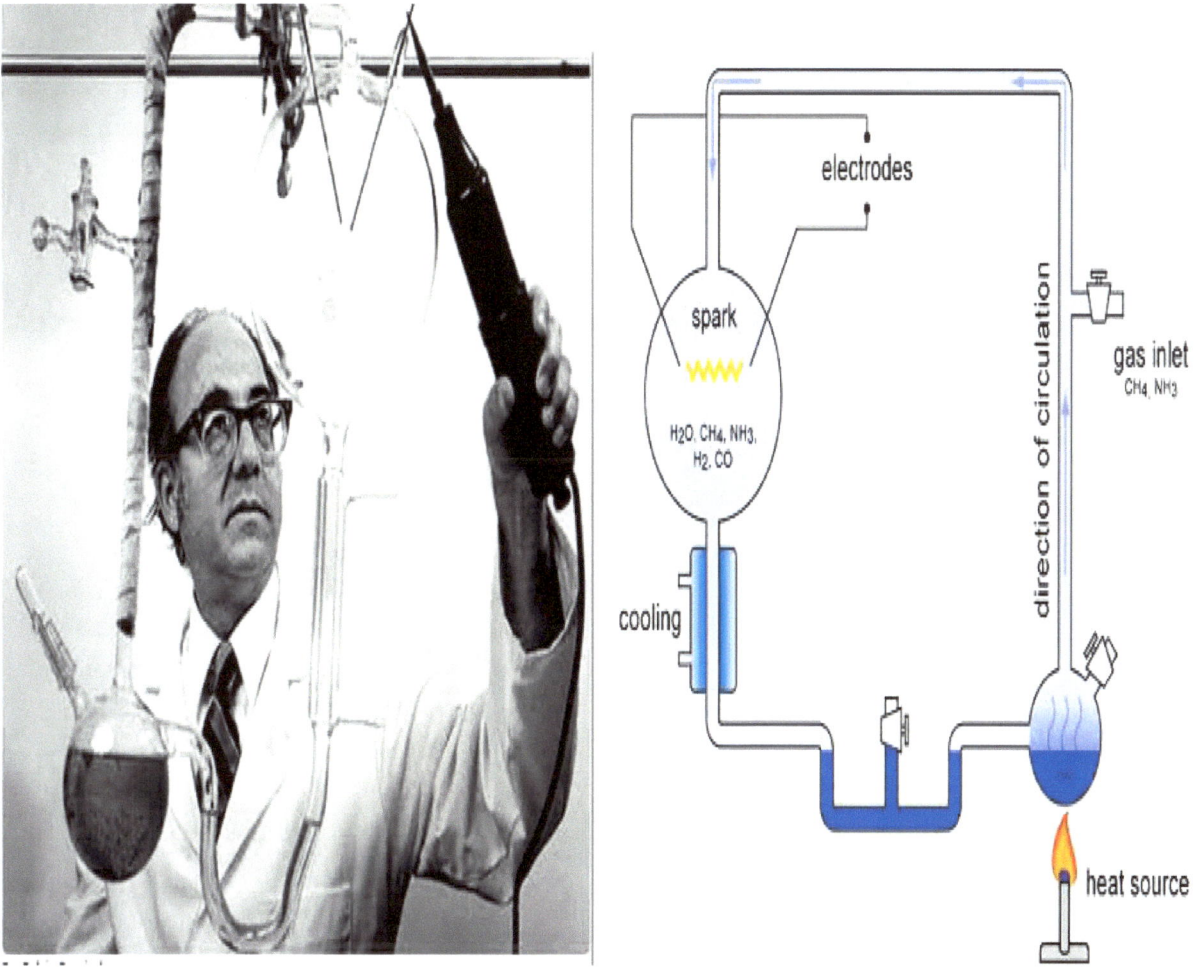

Figure 2. Stanley Miller (Photo Credit: Flickr)

3 Harold Clayton Urey:

Harold Clayton Urey was born on April 29, 1893 in Walkerton, Indiana, United States. He was Nobel Prize Winner in Chemistry (1934). He discovered the deuterium. He contributing the Miller-Urey experiment.

4 Oparin and Haldane Theory:

Life arise from inorganic molecules. Oparin and Haldane thought that, inorganic molecules are reacted to produce amino acids and nucleotides. The first life arose from Ocean's Water.

5 Coacervate:

Coacervates mean aggregates of proteins, amino acids, and other hydrocarbons.

6 Primordial Soup:

The Primordial Soup Theory suggests that chemicals are combined to from amino acids. Amino acids are the building blocks of life. This Theory also suggests that life began on Earth Ocean Water at 3.8 billion years ago.

1.7 Reddi experiment:

We do not know when life originate on the Earth. Most scientists believed that life originate from non-living things such as maggots arose from wheat, lice from sweat, frog from dump mud, fish arose from mud. Several experiment have been conducted to disprove spontaneous generation.

Francesco Reddi is an Italian scientist. Francesco Reddi used to two different jars to this experiment. Both jars were exposed to the atmosphere. One jar was left open while another jar was covered with a cloth. Few days later, he observed the open jar contains maggots whereas the covered jar contains no maggots. Reddi concluded that maggots arose from eggs of the flies. He disproved the spontaneous generation.

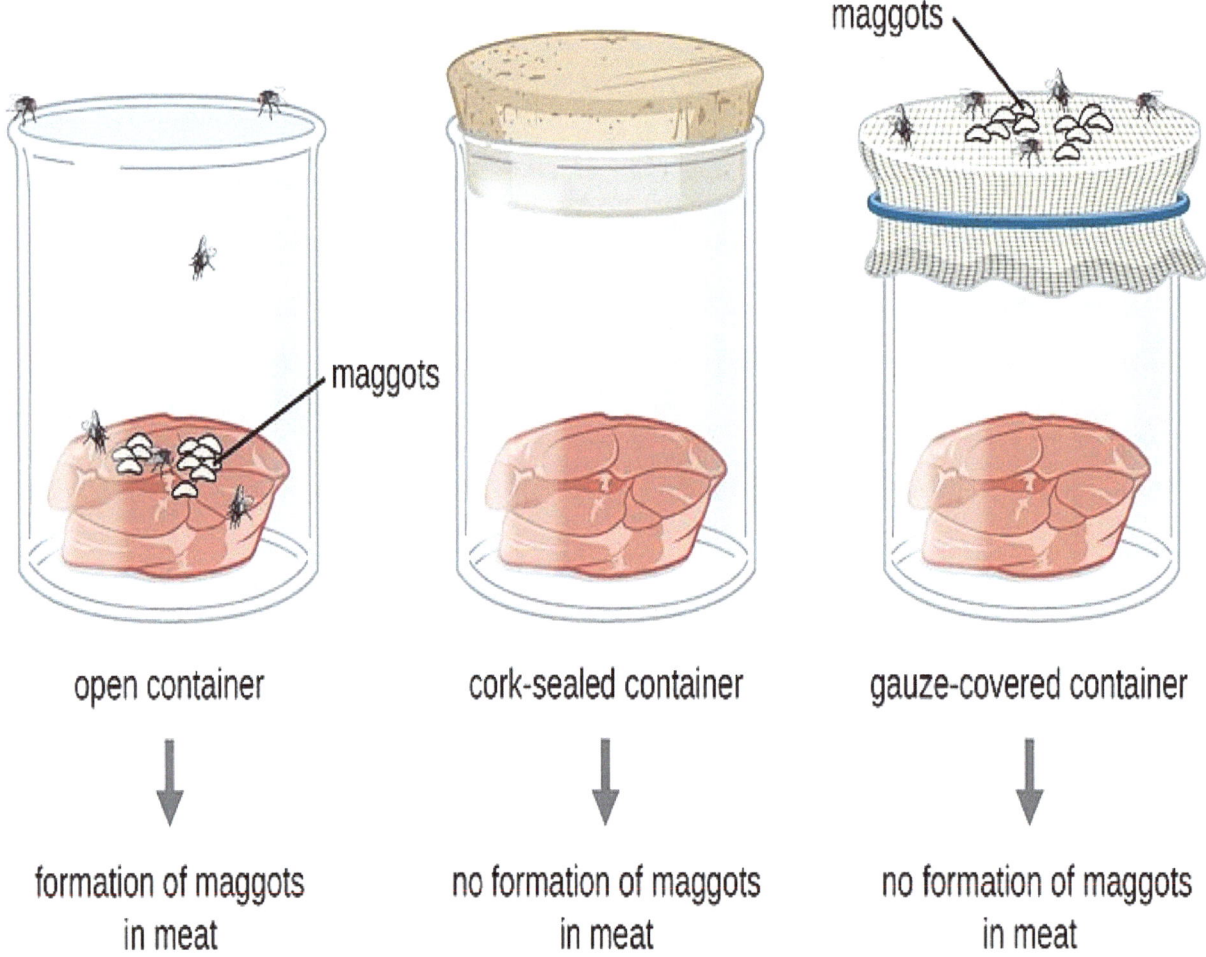

Figure 3. Reddi Experiment (Photo Credit: Wikimedia Commons/CNX OpenStax [CC BY (https://creativecommons.org/licenses/by/4.0)])

1.8 Lazzaro Spallanzani:

Lazzaro Spallanzani was an Italian scientists. Lazzaro spallanzani used four flasks for this experiment. Flask 1 was open and flask 2 was boiled and left open while flask 3 was sealed and flask 4 was boiled and sealed. After this experiment, he observed that falsk 1 and flask 2 was turn cloudy and microbes arises whereas flask 3 turn cloudy and microbes were found and flask 4 was did not turn cloudy and microbes not found.

1.9 Discovery of Microorganisms:

Earth is 4.6 billion years old. Scientist have evident that cells first appeared on Earth between 3.8 billion years ago; these organism were exclusively microbial. In fact microorganisms were the only life on Earth for most of its history. The discovery of the microbial world immediately raised question regarding the origin of microorganisms. Living organisms such as plants and animals do not originate spontaneously. However some belive that these microorganisms arose spontaneously and this theory came to be known as the theory of spontaneous generation or abiogenesis. The basic idea of spontaneous generation can easily be understood. For example, if food is allowed to stand for some time, it putrefies. When the putrefied material examined microscopically, it is found to be teeming with bacteria. Some people said they arose spontaneously from nonliving materials that is spontaneous generation. So, people had belived in generation- that living organism could develop from nonliving matter. But in 1748, the English priest John Needham (1713-1781) reported the results of his experiment on spontaneous generation. needham boiled mutton broth in fused with plant or animal matter, hoping to kill all pre-existing microbes. He then sealed the flask. After a few days, Needham observed that the broth had become cloudy and contained microorganisms. He argued that the new microbes must have arisen spontaneously. Needham (1748) put the question to an experimental test. He wrote: "For my purpose therefore I took a quantity of mutton-gravy hot from the fire, and shut it up in a phial, closed with a cork so well masticated, that my precautions amounted to as much as if I had sealed my phial hermetically. I thus effectually excluded the exterior air, that it may not be said my moving bodies drew their origin from insects, or eggs floating in the atmosphere. I would not instill any water, lest, without giving it as intense a degree of heat, it might be thought these productions were conveyed through that element. My phials swarmed with life." Lazzaro spallanzani (1729-1799) did not agree with Needham conclusions.

Figure 4. Microorganisms (Photo Credit: Flickr)

Today spontaneous generation is generally accepted to have been decisively dispelled during the 19th century by the experiment of Louis Pasteur. Pasteur prepared a nutrient broth similar to the broth one would use in soup. Then, he placed equal amount of the broth into two long nacked flasks. He left one flask with a straight neck. The other he bent to form an S shape. Then he boiled the broth in each flask to kill any living matter in the liquid. The sterile broths were then left to sit, at room temperature and exposed to the air, in their open mouthed flasks. After several weeks, Pasteur observed that the broth in the straight-neck flasks was discoloured and cloudy. Other broth in the curved neck flask had no changed. He concluded that germ in the air were able to fall unobstructed down the straight neck flask and contaminate the broth. Pasteur experiment showed that microbes cannot arise from nonliving materials under the conditions that existed on Earth during his lifetime. But his experiment did not proved that spontaneous generation never occurred. Another contribution of Louis pasteur to germ theory of disease. Germ theory of disease transmission was established by Pasteur. It states tha microorganisms known as pathogens or germs can lead to disease. Germs may refer to any type of microorganism that can cause disease such as protists, fungi, viruses, prions or viriods. Pasteur is regarded the Father of bacteriology and pasteurization.

1.10 Fossil Evidence of Life:

Stromatolites are fossils. Stromatolites are formed by the growth of cyanobacteria.

Figure 5. Stromatolites (Photo Credit: Pixabay)

Figure 6. Stromatolites (Photo Credit: Flickr)

1.11 RNA World hypothesis:

The RNA World Hypothesis was proposed by Carl Woese, Francis Crick, and Leslie Orgel. RNA can also transcribed by reverse transcription.

Alexander Rich was born on November 15, 1924. He was an American Biologist. He first gave the concept RNA World (1962

Walter Gilbert was born on March 21, 1932 in Boston, Massachusetts, United States. He was an American molecular biologist. He first coined the term RNA World in 1986.

Many scientists believe that RNA first come on the Earth because RNA to store genetic information and catalyse chemical reactions.

1.12 What is Life?

There is no definition of life. Millions of years ago, life appears on the Earth. Life requires energy. Life may be reproduce, growth, and metabolise. Life consist of carbon, hydrogen, oxygen, sulphur, phosphorus and hydrogen. We belief that God are creates life to all living organisms. All living organisms are made up of cells. Cells are the fundamentals unit of life. Living thing consume nutrients. Living organisms may be reproduce. Reproduction may be asexual or sexual. Reproduction can create to produce new organism. So, life is capable of reproduction, metabolism, replication, and evolution.

Chapter 2

Biological Macromolecules

Biological macromolecules are carbohydrates, proteins, lipids and nucleic acids.

2.1 Protein:

Proteins are present in all living organisms. Protein consists of amino acid residues joined by peptide bonds. They are composed of one or more chains of amino acids. Amino acids are made up of carbon, hydrogen, nitrogen, oxygen, or sulphur. Amino acids are the building blocks of proteins.

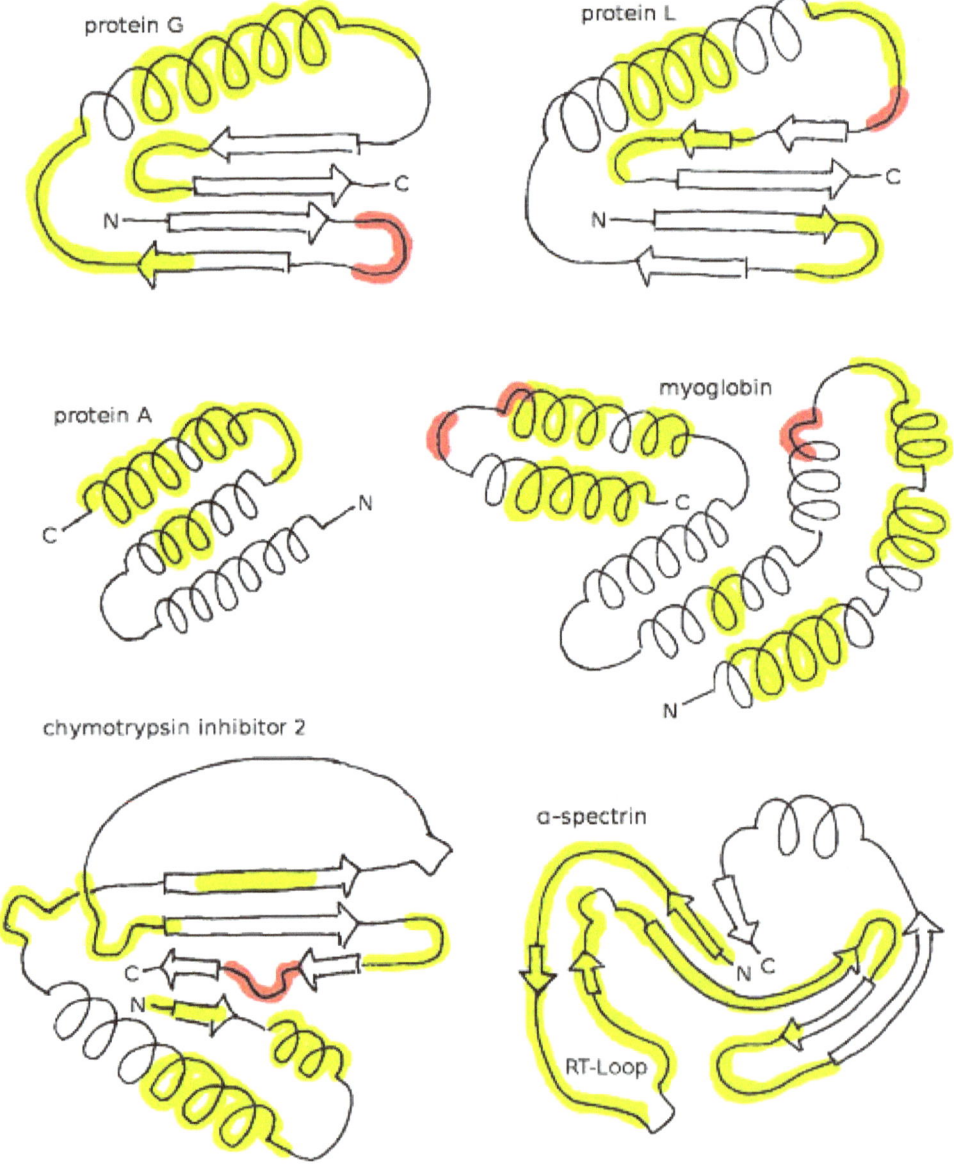

Figure 1. Proteins (Photo Credit: Flickr)

Function of Proteins:

They catalyse metabolic reactions.

They perform essentials for replicating DNA.

Types of Proteins:

Fibrous.

Globular.

Membrane.

Fibrous:

Fibrous proteins are do not denature and generally insoluble in water. They also help in protection and support. Examples of Fibrous proteins are listed below-

- Fibronectin.
- Titin.
- Myosin.
- Spectrin.
- Collagen.
- Tropomysin.
- Elastin.
- Keratin.
- Tau
- Tubulin.

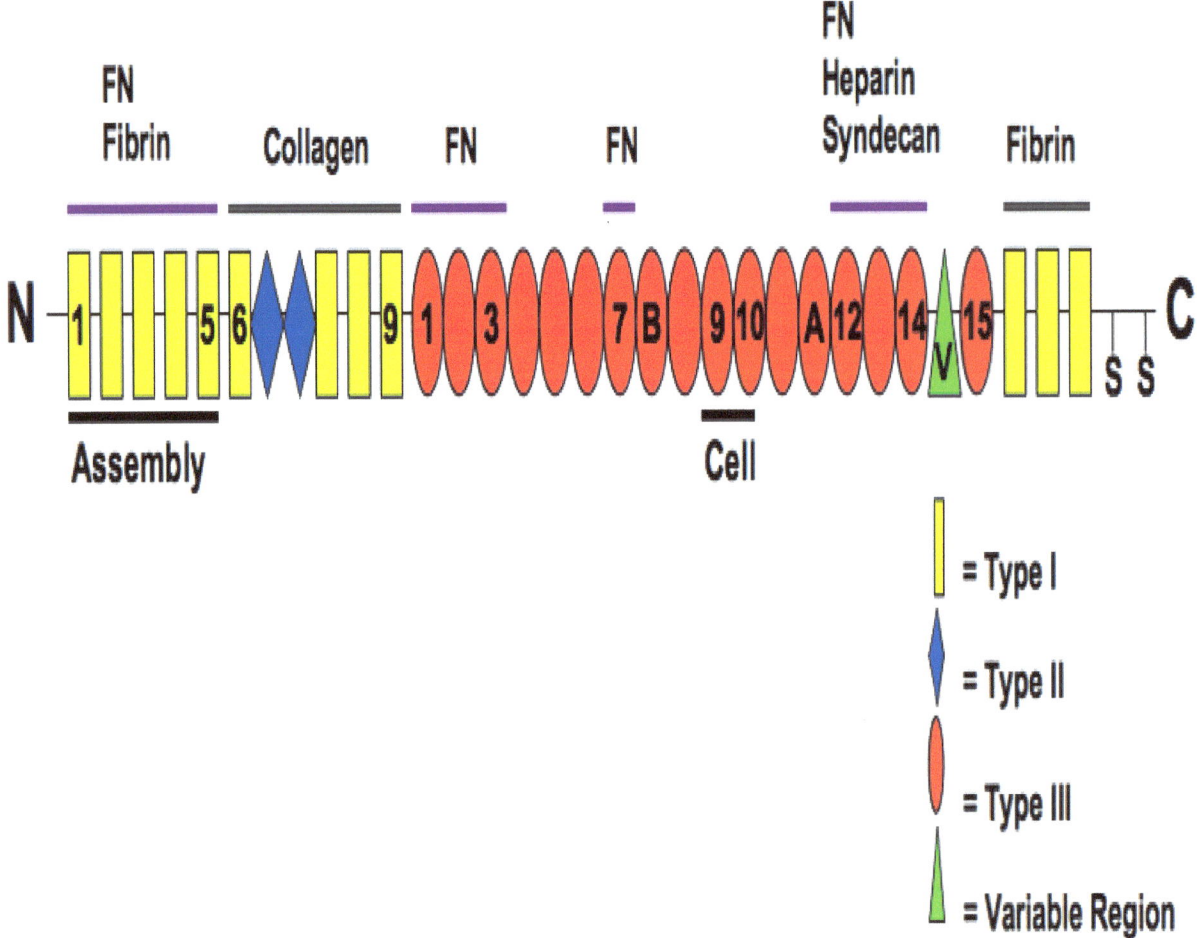

Figure 2. Fibronectin (Photo Credit: Wikimedia Commons/AllWorthLettingGo [CC BY-SA (https://creativecommons.org/licenses/by-sa/3.0)])

Globular Proteins:

Globular proteins are generally water soluble. Globular protein may be round or spherical. This proteins are help in transporting, regulating and catalysing. Examples of Globular proteins are given below-

- Albumins.
- Myoglobin.
- Ependymin.
- Selectin.
- Ig A
- Ig D
- Ig E
- Ig G
- Ig M
- Thrombin.

- Serum Albumin.

Figure 3. IgM (Photo Credit: Wikimedia Commons/ Pecatum [CC BY-SA (https://creativecommons.org/licenses/by-sa/4.0)])

Membrane Protein:

Membrane Proteins are three types including integral membrane protein, peripheral membrane protein, anfd lipd anchored proteins. Examples of membrane proteins are given below-

- CFTR
- Glucose Transported.
- p53

- Rhodopsin.
- Potassium channel.
- Myo D
- Hydrolases.
- Transferases.
- Scramblase.
- C-myc.

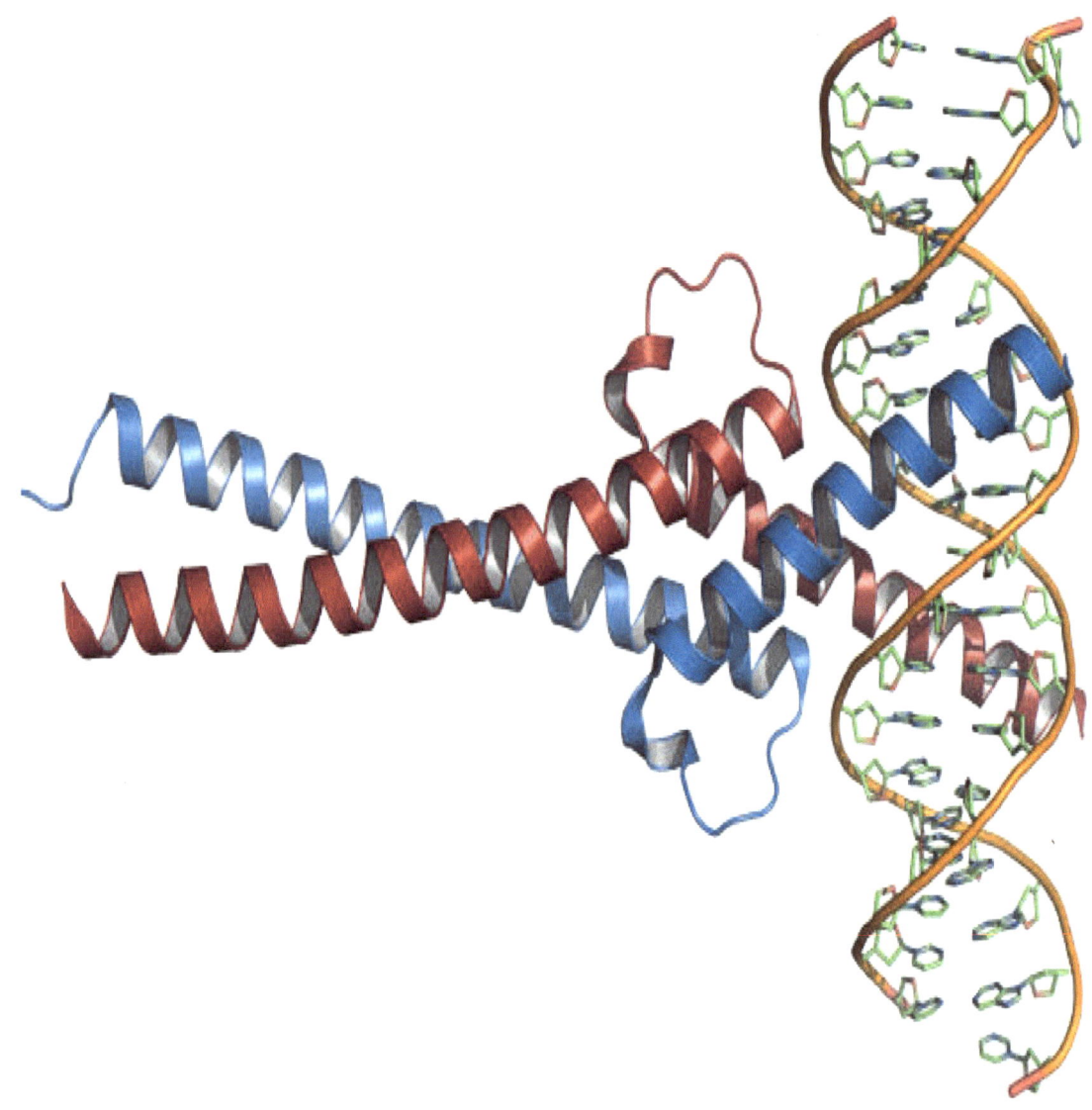

Figure 4. C-Myc (Photo Credit: Wikimedia Commons/ AbsturZ at English Wikipedia [CC BY-SA (https://creativecommons.org/licenses/by-sa/3.0)])

2.2 Primary Structure of Proteins:

Amino acids are link each other to forming proteins. Proteins are the backbone of amino acids. Primary structure of proteins are starting from the (N) terminal end and end to the (C) terminal end. Amino acid contains hydrogen atom, carboxyl group, amino group, and "R" group.

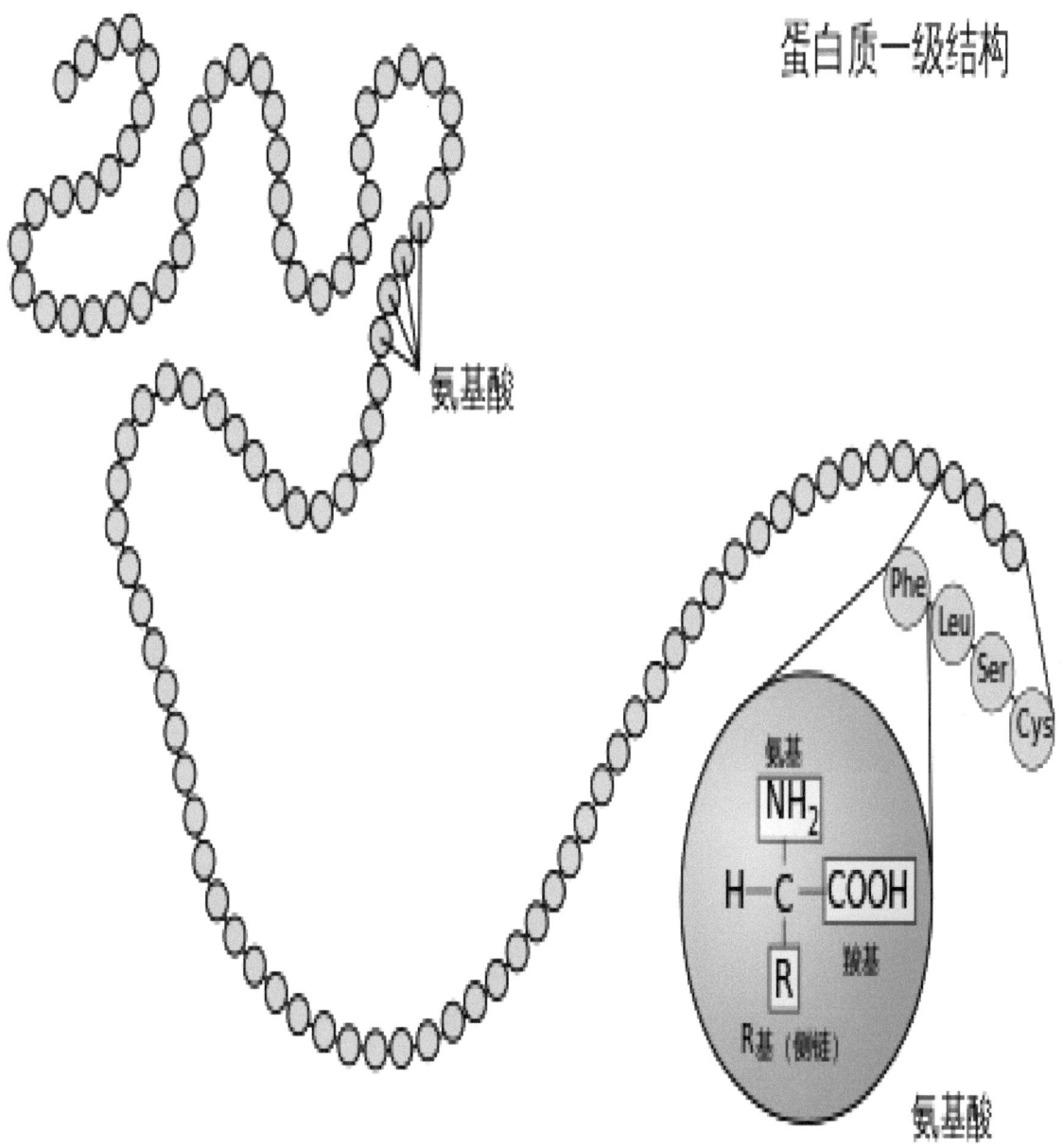

Figure 5. Primary Structure Of Proteins (Photo Credit: Wikimedia Commons/National Human Genome Research Institute, translated into chinese by Webridge [Public domain])

3 Secondary Structure of Proteins:

Secondary Structure of proteins are the α – helix and β- pleated sheet. The α-helix is right hand helix. α- helix are stabilized by hydrogen bonds between the NH and CO groups.

2.4 Tertiary Structure of Proteins:

Secondary structures are converted to form tertiary structure. Tertiary structures are three dimensional shape.

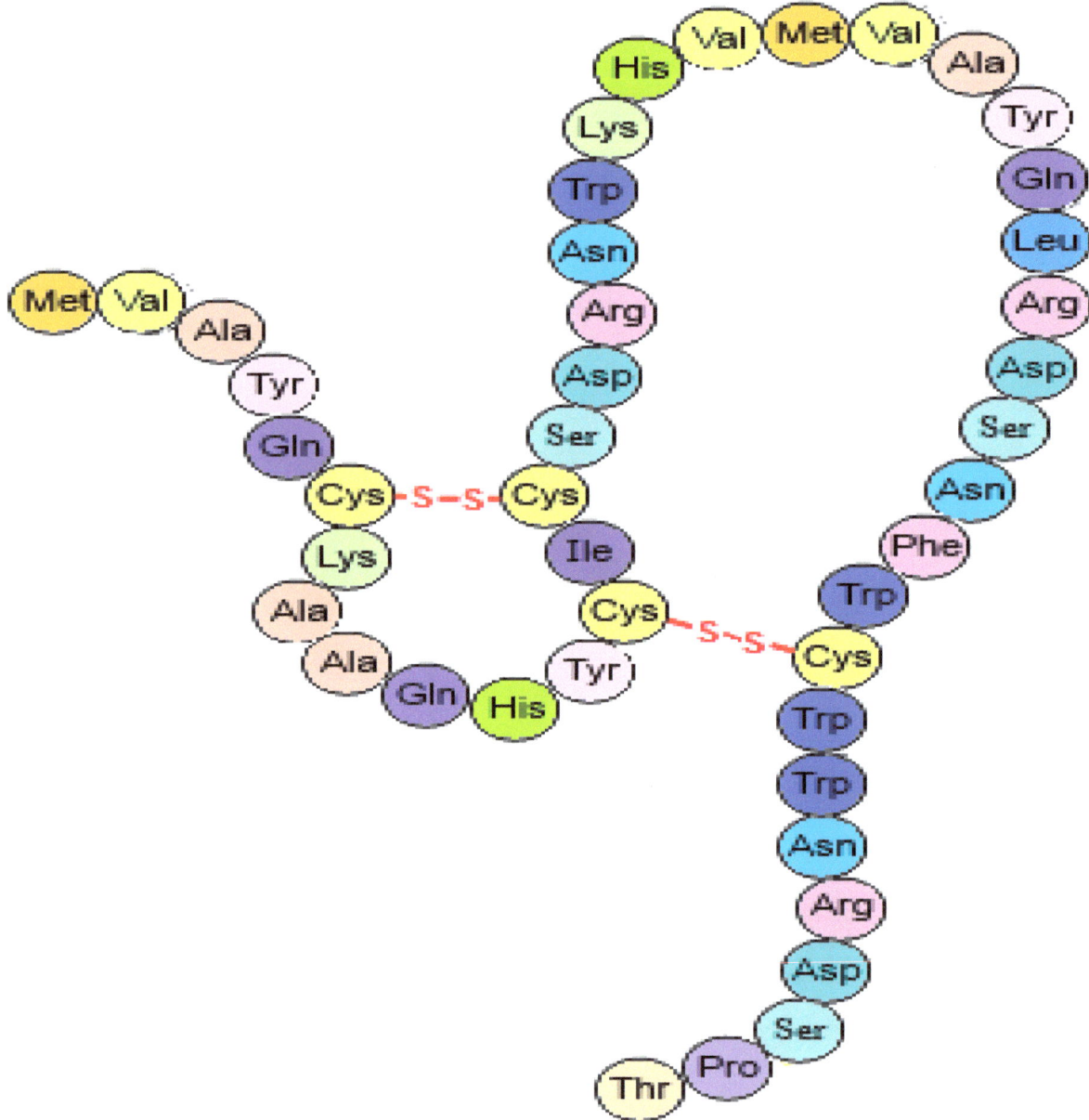

Figure 6. Tertiary Structure of Protein (Photo Credit: Wikimedia Commons/CKRobinson [CC BY-SA (https://creativecommons.org/licenses/by-sa/4.0)])

2.5 Carbohydrate:

Carbohydrates are consists of carbon, hydrogen, and oxygen. Carbohydrates are macronutrients. Carbohydrates are source of energy for the body. Carbohydrates are the building blocks of polysaccharides.

Figure 7. Carbohydrates (Photo Credit: Pixabay)

Types:

Carbohydrates are various types include- monosaccharide, disaccharide, and polysaccharide.

Monosaccharide:

Monosaccharides are the smallest sugar unit. Monosaccharides are colourless and dissolved in water. Glucose, Fructose and Galactose are monosaccharide.

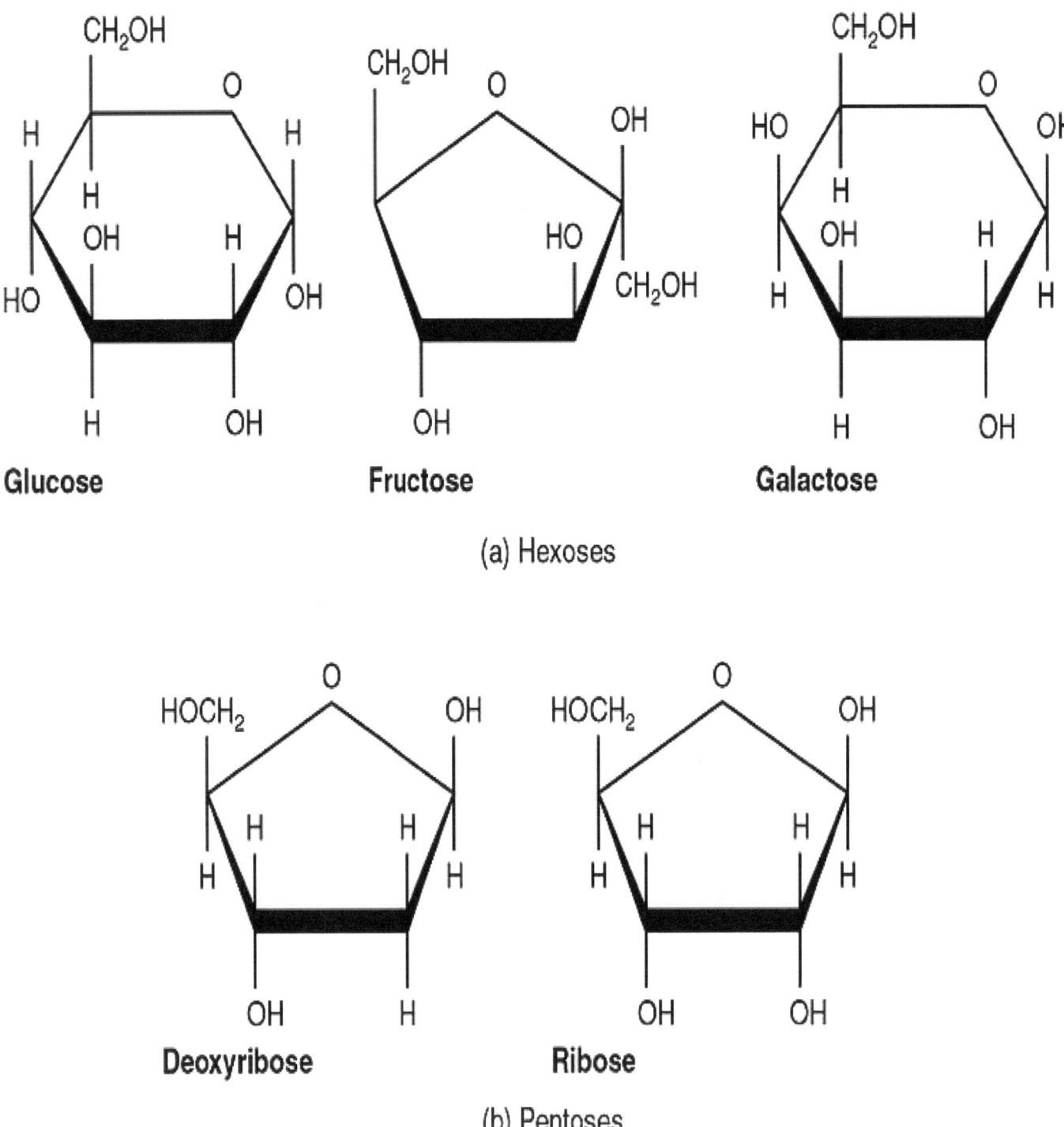

Figure 8. Structure of Monosaccharide (Photo Credit: Wikimedia Commons/OpenStax College [CC BY (https://creativecommons.org/licenses/by/3.0)])

Disaccharide:

When one monosaccharide are joined to another monosaccharide by glycosidic linkage to form disaccharides. Sucrose, maltose, and lactose are disaccharide.

Polysaccharide:

When many monosaccharides are joined to each other forming polysaccharide. Cellulose, Chitin, glycogen, and starch are polysaccharide.

Function of Carbohydrate:

i. Carbohydrate provide energy.

ii. Carbohydrate regulate the blood glucose.

iii. Carbohydrates are essentials for brain function.

Cellulose fibers

Cellulose structure

Figure 9. Carbohydrate Functions (Photo Credit: Wikimedia Commons/CNX OpenStax [CC BY https://creativecommons.org/licenses/by/4.0)])

2.6 Fattty Acids:

Fats are essentials part for health. We cannot live without fat. Fats are non-soluble in water. Fatty acids are the building blocks of fats. Fatty acids are soluble in water.

Types of Fatty Acids:

Fatty Acids are of two types-

i. Saturated Fatty Acids.

ii. Unsaturated Fatty Acids.

Saturated Fatty Acid:

Saturated Fatty acids do not contain double bonds.

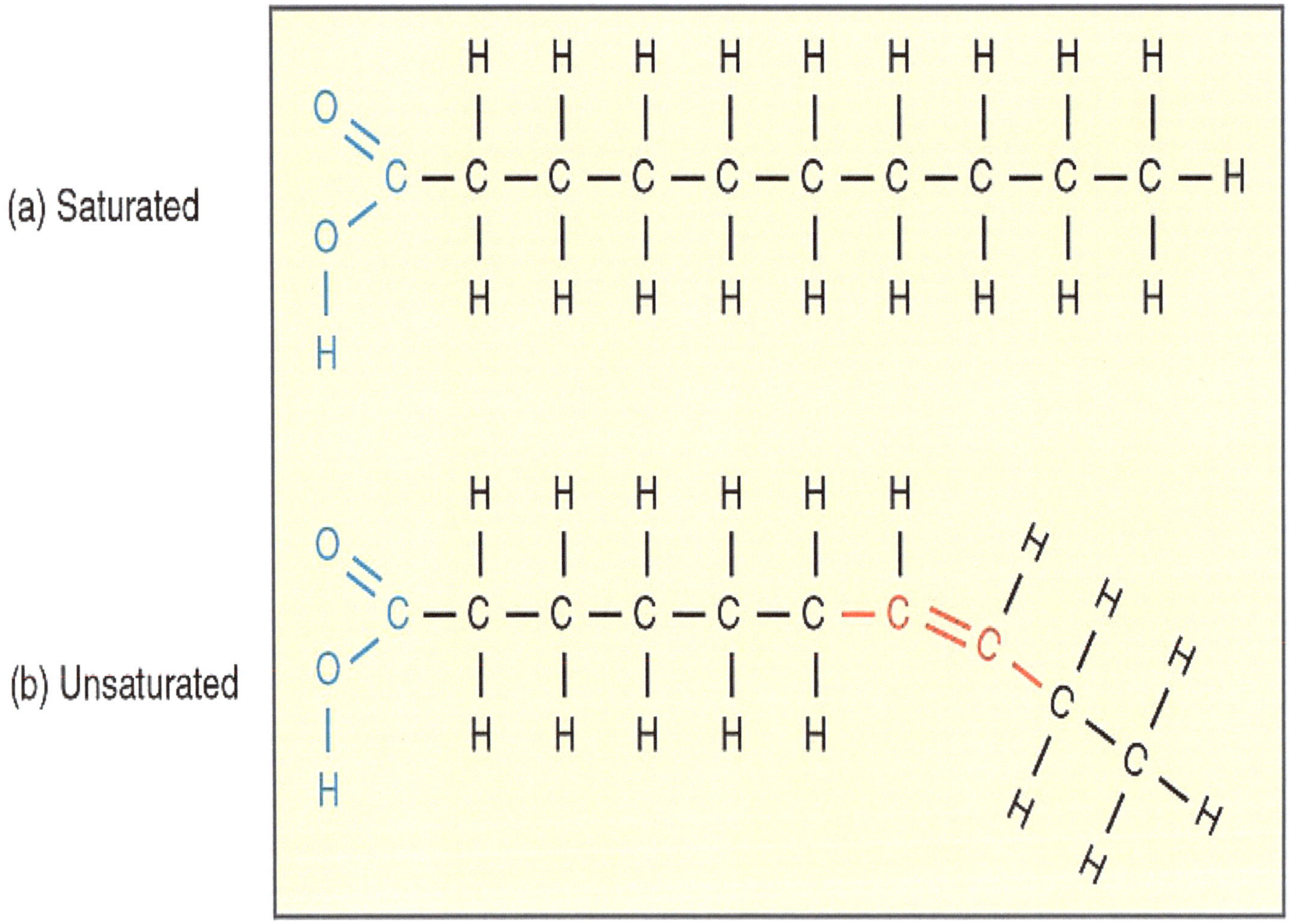

Figure 10. Saturated and Unsaturated Fatty Acids (Photo Credit: Wikimedia Commons/ OpenStax College [CC BY (https://creativecommons.org/licenses/by/3.0)])

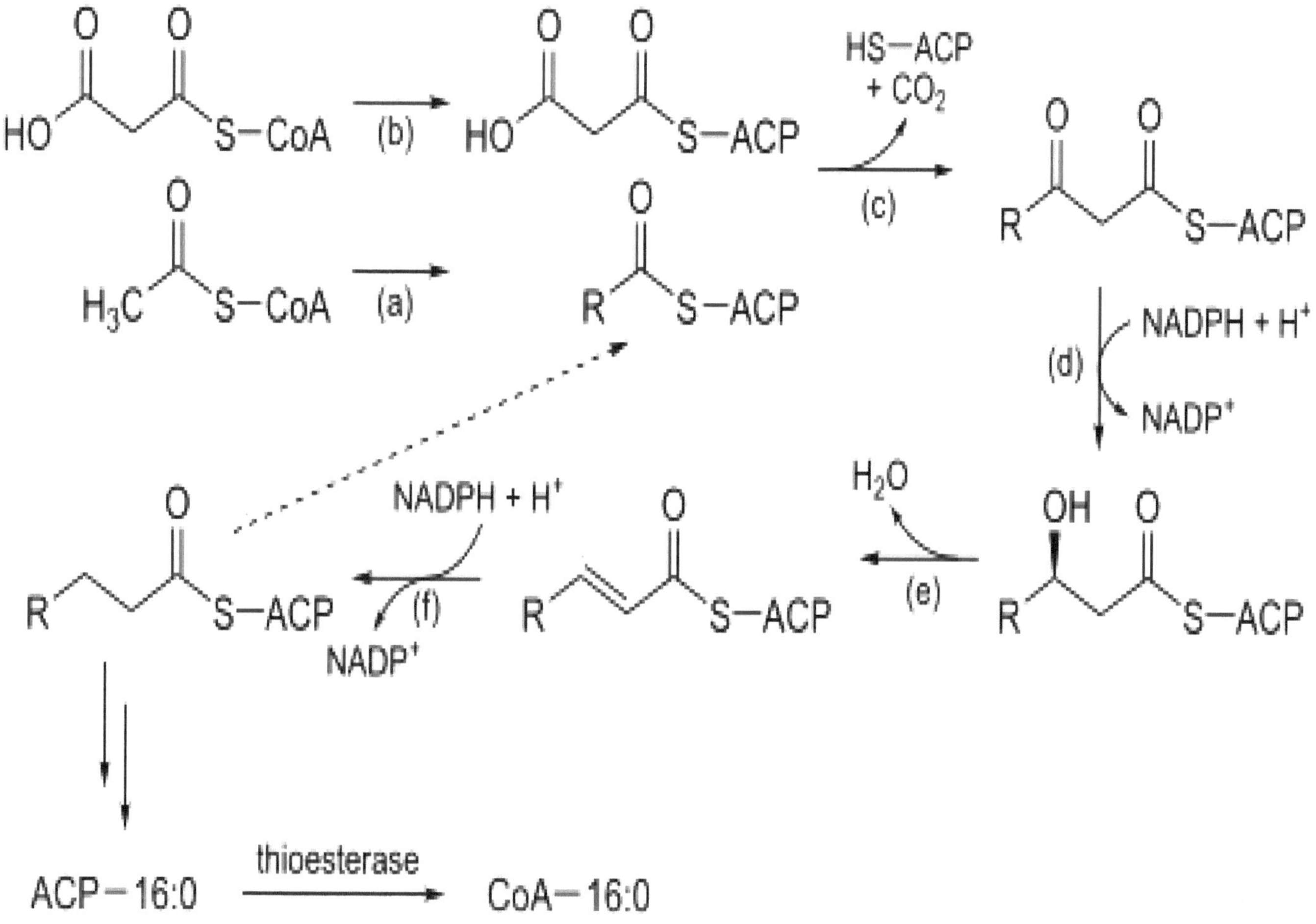

Figure 11. Saturated Fatty Acids (Photo Credit: Wikimedia Commons/Hbf878 [CC0])

Unsaturated Fatty Acids:

Unsaturated Fatty acids contain one or more double bonds.

2.6 Nucleic Acid:

Nucleic acids are building blocks of living organisms. Nucleic acids are macromolecules that transfers genetic information from one generation to another generation. Nucleic acids include deoxyribonucleic acid (DNA) and ribonucleic acid (RNA).

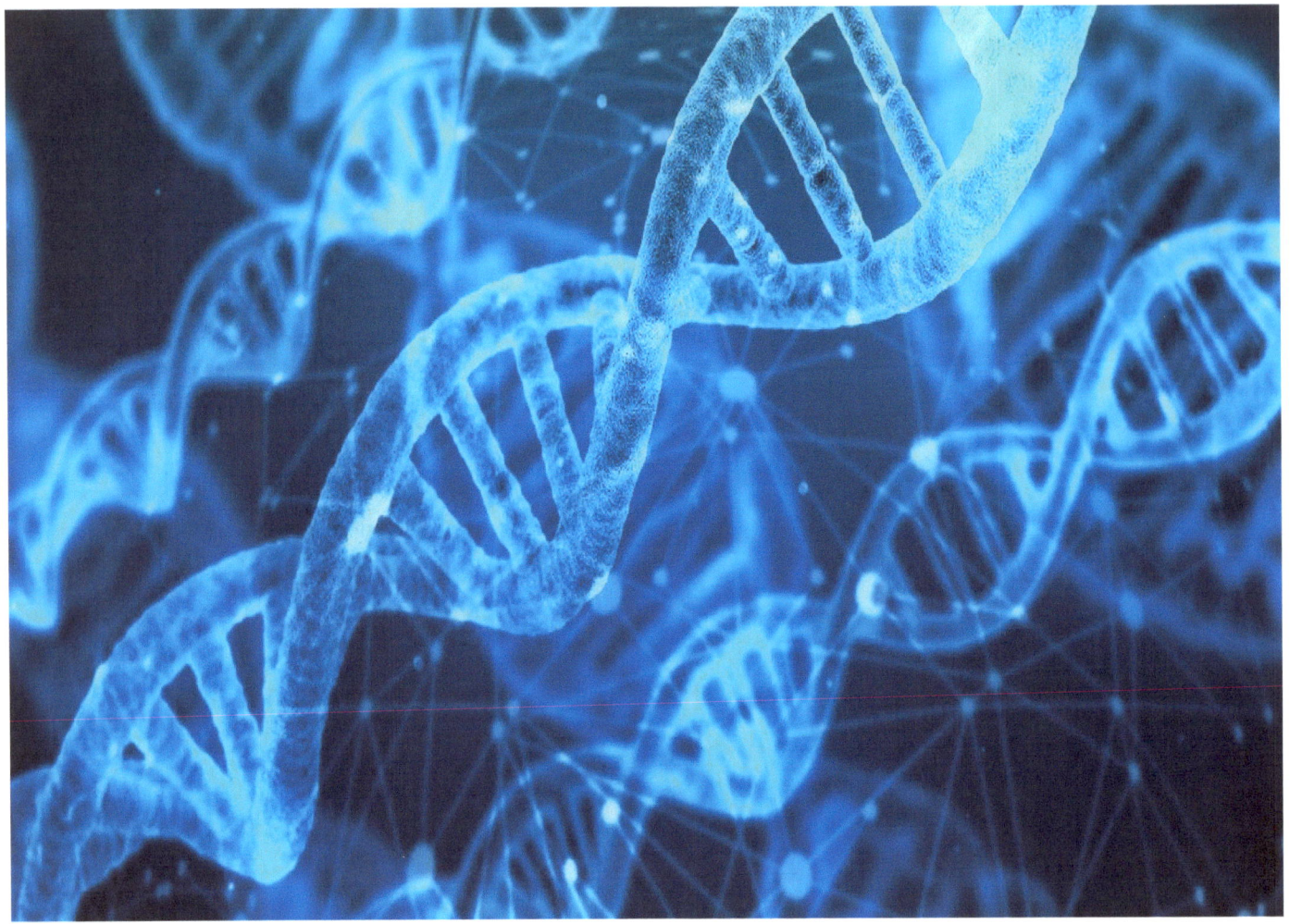

Figure 12. Nucleic Acids (Photo Credit: Pixabay)

Nucleotides are consists are of three parts-

i. Nitrogenous base.

ii. A five carbon sugar.

iii. A phosphate group.

DNA:

DNA is double-stranded and composed of Adenine, Guanine, Cytosine, and Thymine.

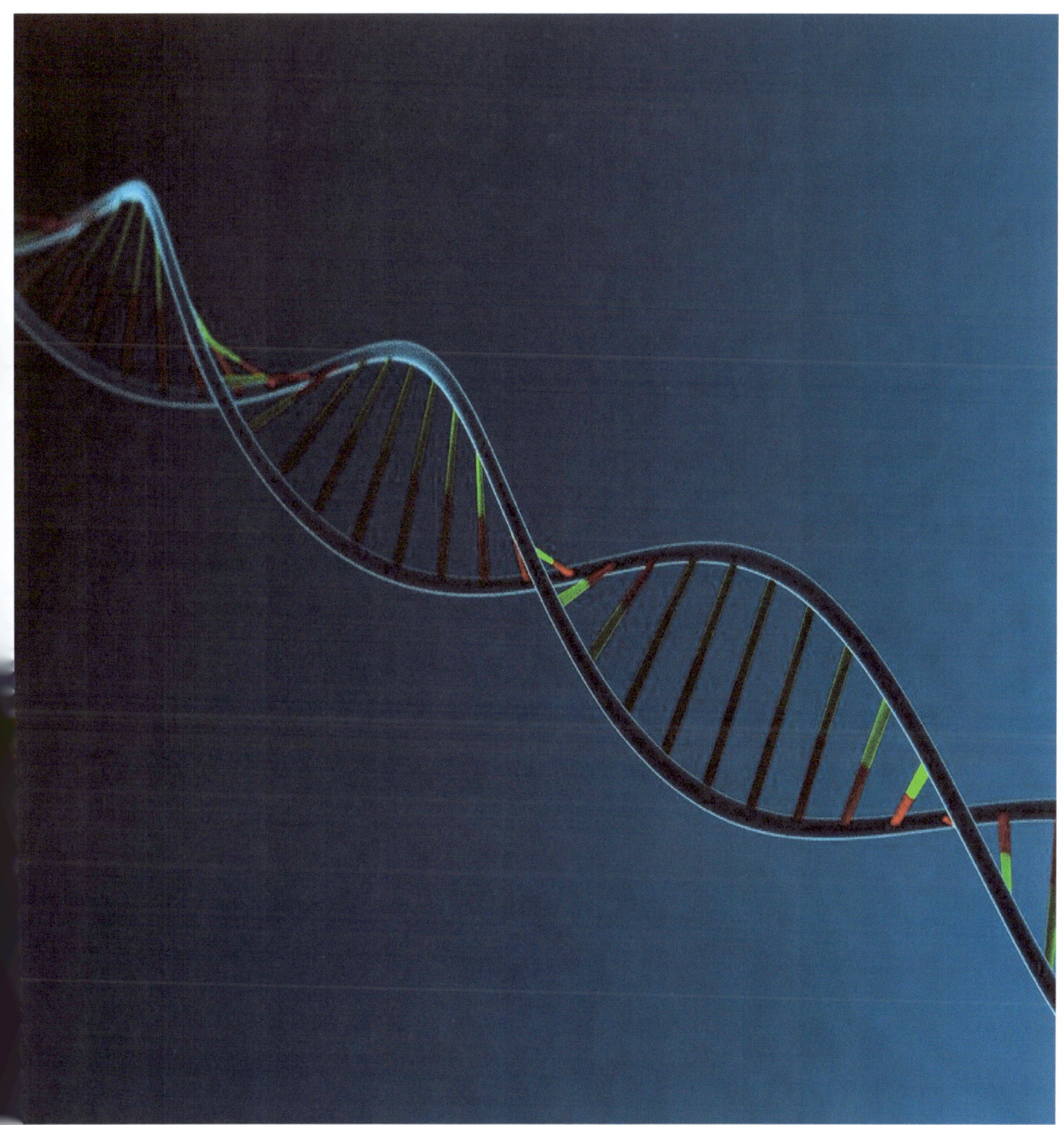

Figure 13. Deoxyribonucleic Acid-DNA (Photo Credit: Pixabay)

RNA:

RNA is single stranded and composed of Adenine, Guanine, Cytosine, and Uracil.

7 Water:

Hydrogen and oxygen mixed with each other to from water. Wter as a solvent. Water is the essential part of our life. Wter are found all over the world. Water molecules are composed of two hydrogen atom linked on oxygen atom. Its chemical formula is H_2O. Water covers 71% of the Earth. Earth surface contains more amounts of water.

Figure 14. Water (Photo Credit: Pixabay)

Traces of water are found on Mercury, Venus, Mars, and The Moon.

References:

1. Miller-Urey Experiment - Amino Acids & Origins of Life on Earth, www.juliantrubin.com/bigten/miller_urey_experiment.html.

2. "A Brief Explanation Of Miller Urey Experiment". BYJUS, 2020, https://byjus.com/biology/miller-urey-experiment/.

3. "Miller–Urey Experiment". En.Wikipedia.Org, 2020, https://en.wikipedia.org/wiki/Miller–Urey_experiment.

4. http://phoenix.lpl.arizona.edu/mars145.php

5. https://study.com/academy/lesson/stanley-miller-theory-experiment-apparatus.html

6. https://en.wikipedia.org/wiki/Stanley_Miller#cite_note-23

7. Reece, Jane B., and Neil A. Campbell. Campbell Biology. Harlow: Pearson Education, 2011. Print.

8. IBM TJ Watson Researcher Center Isidore Rigoutsos Manager Bioinformatics Group, Gregory Stephanopoulos Professor of Chemical Engineering and Biotechnology MIT (2006), Systems.

9. Biology: Volume I: Genomics, Oxford University Press, p. 6.

10. "Origin Of Life: Twentieth Century Landmarks." Origin Of Life: Oparin-Haldane Hypothesis. N.p., n.d. Web. 28 Oct. 2012. <http://www.simsoup.info/Origin_Landmarks_Oparin_Haldane.html>.

11. "The Miller/Urey Experiment." The Miller/Urey Experiment. N.p., n.d. Web. 28 Oct. 2012. <http://www.chem.duke.edu/~jds/cruise_chem/Exobiology/miller.html>.

12. Reece, Jane B., and Neil A. Campbell. Campbell Biology. Harlow: Pearson Education, 2011. Print.

About the Author

Anupam Rajak received his B.Sc in Botany from the Raghunathpur College, SidhoKanho-Birsha University. He has published severals articles in international journal. His email address is anupamrajak1234@gmail.com

www.ingramcontent.com/pod-product-compliance
Lightning Source LLC
Chambersburg PA
CBHW051833210526
45473CB00005B/1860